As Above, So Below
As Within, So Without

The Divine Web: Christ Consciousness and the Universal Neural Network

Steel Waters

Preface

"If you want to find the secrets of the universe, think in terms of energy, frequency, and vibration." -Nikola Tesla

In the quest for deeper understanding of existence, Nikola Tesla's words, "If you want to find the secrets of the universe, think in terms of energy, frequency, and vibration," reveal a profound truth that aligns with the essence of Christ Consciousness. This divine awareness embodies the energetic fabric of love and unity that permeates all creation. As we explore Christ Consciousness, we discover the interconnectedness of all things—a vibrational field where spirit and matter coexist. This book delves into these universal principles, uncovering how Christ Consciousness manifests through energy systems, healing practices, and the spiritual science of the cosmos. Its purpose isn't to reveal all of the secrets of the Universe, but to provide a guide on how to reveal the Universe inside of each one of us.

We will examine how the recognition of interconnectedness, oneness, and divine unity lies at the heart of both spiritual principles, offering insights into the nature of reality and the path to spiritual awakening.

Contents

1. Introduction to Christ Consciousness and Universal Interconnectedness
- What is Christ Consciousness?
  - Definition, historical context, and significance of Christ Consciousness.
  - Universal love, unity, and the divine within all beings.
- Christ Consciousness and the Universe
  - The universe as a living, interconnected entity.
  - Quantum theory and spirituality: exploring the overlap between science and metaphysics.
  - The universe as a giant neural network and Christ Consciousness as the connecting thread.

2. The Ether: The Fifth Element
- The Ether in Ancient and Modern Thought
  - The role of ether as the "fifth element" in various ancient cultures (Greek, Egyptian, Indian).
  - Ether as the medium that connects all existence—physical, mental, spiritual.
- Quantum Field Theory and the Ether
  - Modern interpretations of the ether within quantum field theory and its relationship to Christ Consciousness.

3. Sacred Geometry: The Blueprint
- Sacred Geometry Across Cultures
  - Historical significance of sacred geometry in Egyptian, Hindu, and Mayan cultures.
  - Key patterns (Flower of Life, Fibonacci sequence) as reflections of divine order.
- Christ Consciousness and Sacred Geometry
  - Christ Consciousness as the guiding force behind geometric patterns that reflect the universe's structure.

4. The Chakra System: Energy Vortices of the Human Body
- Understanding the Chakras
  - Overview of the chakra system from Eastern spiritual traditions.
  - Connection between chakras and physical, emotional, and spiritual well-being.
- Chakras and Christ Consciousness
  - How Christ Consciousness flows through and aligns with the chakra system for spiritual awakening.

5. Reiki: Healing Through Universal Life Force
- The Origins and Principles of Reiki
  - Reiki as a Japanese energy healing system using the universal life force.
  - Reiki and Christ Consciousness
  - How Reiki channels Christ Consciousness for healing and attuning to divine energy.

6. The Human Electromagnetic Field (EMF) as a Manifestation of Christ Consciousness
- Human EMF and Its Role in Consciousness

- The scientific basis for the human EMF and its link to health and consciousness.
- EMF and Christ Consciousness
  - Strengthening the human EMF through spiritual practices to enhance connection with divine energy.

7. The Electromagnetic Field of the Natural Environment
- Earth's Electromagnetic Field
  - Exploration of the Earth's magnetic field as a conscious, living system.
- Human-Environment Connection
  - Interaction between human EMFs and the Earth's EMF, fostering harmony between the individual and nature.

8. The Power of Sound: Vibrations, Healing, and Christ Consciousness
- Sound Healing Across Cultures
  - Examination of sound healing practices in Tibetan, Indigenous, and other spiritual traditions.
- Christ Consciousness and Sound
  - Sound as a carrier of divine energy and Christ Consciousness.

9. Alternative Medicines from Around the World: Christ Consciousness and Healing
- Global Healing Traditions
  - Exploration of Ayurveda, Traditional Chinese Medicine, Indigenous healing practices, and their common themes of balance and energy.
- Christ Consciousness and Healing
  - Aligning with Christ Consciousness for holistic healing across physical, emotional, and spiritual dimensions.

10. The Journey to Awakening: Embodying Christ Consciousness in Everyday Life
- Daily Practices for Awakening Christ Consciousness
  - Meditation, breathwork, energy healing, journaling, and acts of service as daily practices.
- Living in Alignment with Universal Consciousness
  - Living mindfully and in harmony with the universe's energetic patterns.

Chapter 1: Introduction to Christ Consciousness and Universal Interconnectedness

Christ Consciousness refers to a sense of awareness in which one realizes their divine nature and the interconnectedness of all life. This concept transcends any one religion, though it draws heavily from the teachings of Jesus Christ, whose embodiment of divine love, compassion, and unity serve as a model for spiritual awakening. In the Gospels, Christ often speaks of the Kingdom of Heaven being within (Luke 17:21), pointing to an inner realization of divine presence that every individual can awaken to.Historically, Christ Consciousness has evolved as a metaphysical concept. Early Christian mystics, such as St. Augustine and St. John of the Cross, alluded to the idea of a divine spark within every soul. This consciousness of divine unity also mirrors Eastern concepts, such as Buddha Nature in Buddhism or Atman in Hinduism. These terms all reflect the same core idea: that divinity is not external but resides within and can be experienced directly. Christ Consciousness the internal recognition and acknowledgement of the divine unity that binds all creation together. It is not limited to the Christian faith but can be applied universally to any individual who awakens to the understanding that they are part of the greater whole. This consciousness transcends personal identity, ego, and separateness, aligning the individual with the divine flow of love, wisdom, and compassion.

Unity consciousness refers to a state of awareness in which an individual experiences a profound sense of interconnectedness with all of existence. It transcends dualistic thinking and separateness, recognizing the underlying unity that binds all beings and phenomena together. At its core, unity consciousness acknowledges that we are not separate from each other, from nature, or from the divine. It emphasizes the inherent oneness of all life and the recognition of the divine presence within oneself and others. Unity consciousness is central to the teachings of Christ Consciousness, as exemplified by Jesus Christ's teachings on love, compassion, and the kingdom of God. Jesus emphasized the interconnectedness of all beings and the importance of loving one's neighbor as oneself.

In Christ Consciousness, the recognition of unity extends beyond human relationships to include a profound sense of unity with the divine. Jesus' teachings on the Father-Son relationship and the unity of believers reflect this deeper understanding of divine oneness. The concept of Christ as the embodiment of divine love and unity underscores the idea that all individuals have the potential to awaken to their true nature as expressions of the divine. Christ Consciousness invites individuals to realize their inherent unity with God and to embody love, compassion, and unity in their lives.

Embracing unity consciousness has practical implications for how we relate to ourselves, others, and the world around us. It fosters empathy, compassion, and cooperation, leading to greater harmony and peace in our relationships and communities. Unity consciousness also informs our understanding of spirituality and the nature of reality. It invites us to see beyond the illusion of separation and to recognize the interconnected web of life that sustains us all. Practicing unity consciousness involves cultivating qualities such as empathy, compassion, gratitude, and forgiveness. It encourages us to see the divine presence in everyone we meet and to treat others with kindness and respect.

The Teachings of Jesus on Divine Unity

In the Bible, Jesus emphasizes the divine unity that Christ Consciousness points to. In John 10:30, He declares, "I and the Father are one," suggesting the non-dual nature of Christ Consciousness. This statement signifies the realization that God and His creation are not separate but interconnected. Jesus also encourages His followers to love others as themselves (Mark 12:31), which is the practical outworking of Christ Consciousness—recognizing the inherent oneness in all beings.

Christ Consciousness and the Universe

To grasp Christ Consciousness fully, one must expand beyond the personal, seeing the universe itself as a living, conscious entity. The idea that the universe is interconnected and conscious aligns with several spiritual and scientific views. From a quantum perspective, the universe is not a random, mechanistic system but a dynamic, interconnected web of energy. Quantum theory, especially the concept of quantum entanglement, suggests that particles, no matter how far apart, remain connected, impacting each other instantly. This mirrors the idea of divine unity found in Christ Consciousness, where all beings and phenomena are interconnected. As physicist David Bohm described it, the universe operates more like a "holistic order," an undivided whole in which the parts are enfolded. In the realm of spirituality, the overlap with quantum theory offers profound insights. Many mystics and modern spiritual teachers, such as Eckhart Tolle and Deepak Chopra, have likened spiritual awakening to understanding this deep interconnectedness at a quantum level. Christ Consciousness suggests that this network of connection is not just physical but also deeply spiritual—woven through divine intelligence. One intriguing model for understanding Christ Consciousness is the comparison of the universe to a giant neural network. Just as the human brain operates through billions of interconnected neurons, transmitting information through electrical impulses, the universe too can be seen as a vast, interconnected web of energy and consciousness.

Neural networks in the brain are not isolated; they operate in harmony, communicating with one another to form thought, sensation, and awareness. Similarly, at a cosmic level, every part of the universe is interconnected, exchanging energy and information, creating a living, breathing neural network on a universal scale. The structure of the universe exhibits remarkable parallels with the structure of neural networks. Research from neuroscientists and astrophysicists has shown that the distribution of matter in the universe—galaxies, dark matter, etc.—bears a striking resemblance to the neural pathways of the brain. Both systems are vast, complex, and interconnected. This model suggests that Christ Consciousness operates like the neural impulses in the brain, transmitting divine energy and wisdom throughout the universe.

Christ Consciousness can be seen as the guiding intelligence or divine energy that flows through this universal neural network. As with neurons communicating through electrical signals, Christ Consciousness is the divine frequency that flows through all life, connecting every individual with the divine source. In the Bible, Jesus states, *"I am the vine; you are the branches" (John 15:5),* reflecting this same interconnected relationship. Just as neurons in a brain connect to form a unified consciousness, all of creation is connected through Christ Consciousness, forming one universal mind.

Biblical References

1. John 10:30 – "I and the Father are one."
2. Luke 17:21 – "The Kingdom of God is within you."
3. John 14:20 – "On that day you will realize that I am in my Father, and you are in me, and I am in you."
4. John 15:5 – "I am the vine; you are the branches."
5. 1 Corinthians 3:16 – "Don't you know that you yourselves are God's temple and that God's Spirit dwells in your midst?"
6. John 17:21 – "That all of them may be one, Father, just as you are in me and I am in you."
7. Romans 12:5 – "So we, being many, are one body in Christ, and every one members one of another."
8. Ephesians 4:4-6 – "There is one body and one Spirit... one God and Father of all, who is over all and through all and in all."
9. Colossians 3:11 – "Christ is all, and is in all."
10. John 8:12 – "I am the light of the world. Whoever follows me will never walk in darkness, but will have the light of life."

Chapter 2:
Ether-The Divine Medium

In Chapter 1, we explored the idea of the universe as a giant neural network. The ether, or quantum field, can be seen as the medium through which this cosmic neural network operates. Just as neurons in the brain communicate through electrical impulses traveling across synapses, the universe communicates through etheric vibrations that carry divine intelligence and Christ Consciousness. Through these vibrations, every part of the universe is connected, just as the neurons in the brain form a cohesive, unified system. By tapping into the ether, individuals can connect to this divine intelligence and align themselves with the flow of Christ Consciousness The ether (also known as aether or quintessence) is an ancient concept that has persisted across various cultures and spiritual traditions. It is often regarded as the fifth element, complementing earth, air, fire, and water, and considered the invisible substance that fills all space and connects all things. In ancient civilizations, the ether was believed to be the medium through which divine forces manifested and through which all life was interconnected.

Historical Views on Ether-Ether as the "Fifth Element"

In each of these traditions, ether was regarded as the most subtle and refined element, transcending the material world. It was viewed as the divine medium through which higher consciousness and spiritual energies operated. Modern spiritual teachings often equate the ether with the spiritual plane, the realm beyond physical reality, where consciousness and divine intelligence reside.

- Ancient Greece: The philosopher Aristotle referred to ether as the divine substance that made up the celestial bodies and the heavens. He described it as a perfect, unchangeable, and luminous element, unlike the other four classical elements that made up the earthly realm.

- Ancient Egypt: The Egyptians believed that Akasha, or ether, was the life force that permeated the universe. It was considered the breath of the gods, the substance from which creation itself arose.

- India (Vedic Tradition): In Hindu cosmology, ether is known as Akasha, one of the five elements (Pancha Bhutas). Akasha is the infinite, boundless space where all other elements exist. It is associated with the sky, the cosmos, and spiritual energy, and it serves as the medium through which vibrations of sound, light, and consciousness travel.

- Medieval Alchemy: Ether, or quintessence, was central to the alchemists' search for the philosopher's stone. They believed it to be the pure essence that could bring transformation and enlightenment. Ether was seen as the divine matter that held the potential for spiritual alchemy, enabling practitioners to transmute base metals into gold and attain immortality.

The Ether and Christ Consciousness

In the context of Christ Consciousness, ether can be understood as the medium through which divine love, unity, and wisdom flow throughout the universe. Christ Consciousness is often described as the awareness of the divine unity of all creation, and the ether is the conduit for that awareness.

As the divine substance that permeates all space, ether serves as the bridge between the physical and the spiritual realms. It is the medium through which Christ Consciousness can be experienced and expressed in the material world. Jesus' teachings often emphasize the idea of oneness with God and the universe, and ether is the metaphysical space where this unity is realized. For example, when Jesus speaks of the Kingdom of Heaven being "within you" (Luke 17:21), He may be pointing to the idea that divine consciousness flows through the ether, connecting all beings. By attuning oneself to the ether, one can become more aware of this divine presence, experiencing the unity and love that Christ embodied.

Ether is not limited to the physical or spiritual dimensions; it connects all levels of existence. It is the medium through which the body, mind, and spirit interact. In this way, Christ Consciousness—divine awareness—flows seamlessly between these levels, allowing individuals to experience healing, transformation, and spiritual awakening. The concept of ether is particularly important in esoteric traditions, where it is believed that energy flows through the etheric body, a subtle energy field that surrounds the physical body. This etheric body is responsible for channeling life force (or prana) and transmitting it to the physical body, promoting health and well-being. In the same way, Christ Consciousness flows through this etheric field, awakening individuals to their divine nature.

Quantum Field Theory and the Ether

With the advent of modern physics, the notion of ether was largely dismissed in favor of the theory of space-time as described in Einstein's theory of relativity. However, recent developments in quantum field theory have reignited interest in the idea of an underlying medium that permeates all space—an idea that closely parallels the ancient concept of ether.

Quantum field theory proposes that the universe is filled with quantum fields—energy fields that give rise to particles and forces. These fields are constantly fluctuating, even in the vacuum of space. In a sense, these quantum fields act like the modern version of ether, providing the foundation for all matter and energy in the universe. Christ Consciousness, in this framework, can be understood as the divine intelligence or zero-point energy that permeates these fields, infusing them with consciousness and purpose. Just as quantum fields underlie all matter and energy, Christ Consciousness can be seen as the divine intelligence that informs and animates the universe. This consciousness operates at the quantum level, orchestrating the unfolding of reality and guiding the flow of energy and information.

For example, the phenomenon of **quantum entanglement**—where two particles remain interconnected regardless of the distance between them—illustrates the profound unity that exists at the quantum level. In the same way, Christ Consciousness connects all beings, transcending space and time to create a state of divine unity.

Practices for Connecting with the Ether

To attune oneself to the ether and experience Christ Consciousness, spiritual practices that focus on energy flow and vibration are essential. These practices help individuals become more sensitive to the subtle energies of the etheric field and allow them to experience the interconnectedness of all life.

1. Meditation: Deep meditation practices, particularly those focused on breath and sound, can help individuals access the ether. The breath is often seen as the bridge between the physical and spiritual realms, and by focusing on the breath, one can become more aware of the etheric field surrounding and permeating the body.

2. Breathwork: Practices such as **Pranayama** in the yogic tradition or **Holotropic Breathwork** can open the etheric channels and allow the life force (prana) to flow freely. This helps to activate Christ Consciousness by aligning the body, mind, and spirit with divine energy.

3. Energy Healing: Systems such as **Reiki** and **Qi Gong** are based on the flow of etheric energy through the body. By working with the etheric field, these healing practices can clear energy blockages and restore balance, allowing Christ Consciousness to flow more freely through the individual.

Biblical References
1. Acts 17:28 – "For in Him we live and move and have our being."
2. Hebrews 1:3 – "The Son is the radiance of God's glory and the exact representation of His being, sustaining all things by His powerful word."
3. Colossians 1:17 – "He is before all things, and in Him all things hold together."
4. Psalm 139:7-8 – "Where can I go from your Spirit? Where can I flee from your presence? If I go up to the heavens, you are there; if I make my bed in the depths, you are there."
5. Ephesians 4:6 – "One God and Father of all, who is over all and through all and in all."

Chapter 3:
Sacred Geometry- The Blueprint

Sacred geometry is the study of the geometric patterns and shapes that are believed to form the foundation of all creation. These shapes are not merely mathematical structures but are thought to reflect the deeper truths of the universe and consciousness itself. Many ancient civilizations, such as the Egyptians, Greeks, Hindus, and Mayans, regarded these geometric patterns as sacred, encoding the divine principles that govern the physical and spiritual realms.

Historical Significance of Sacred Geometry

- Egyptian Civilization: In ancient Egypt, the construction of pyramids, temples, and monuments incorporated sacred geometric principles. The Great Pyramid of Giza, for example, is based on the golden ratio, also known as phi (ϕ), which reflects the proportions found in natural forms such as plants, seashells, and even galaxies. Egyptians believed that by aligning structures with sacred geometry, they were connecting with divine energies.

- Hinduism: In the Vedic tradition, sacred geometry is reflected in the design of yantras and mandalas. These geometric designs are used for meditation and are believed to represent cosmic forces and spiritual energies. The Sri Yantra, a complex geometric design composed of interlocking triangles, is one of the most well-known and symbolizes the union of masculine and feminine energies, as well as the connection between the individual and the universe.

- Greek Civilization: The ancient Greeks, particularly Pythagoras, viewed geometry as a reflection of the divine order of the cosmos. Pythagoras discovered the mathematical relationships between harmonic sounds and geometric shapes, believing that these relationships revealed the underlying harmony of the universe. The Platonic solids—five three-dimensional shapes in which all faces, edges, and angles are identical—were seen as the building blocks of creation, each corresponding to one of the classical elements: earth, air, fire, water, and ether.

- Mayan Civilization: The Mayans used sacred geometry in their architecture and calendar systems. The design of their pyramids, like the Pyramid of Kukulkan at Chichén Itzá, was based on precise astronomical and geometric calculations, reflecting their deep understanding of cosmic cycles. The Mayans believed that sacred geometry allowed them to align their structures with celestial movements, thereby connecting the physical world with the divine.

Key Shapes in Sacred Geometry

- Flower of Life: The Flower of Life is a complex geometric shape consisting of multiple overlapping circles. It is considered one of the most important symbols in sacred geometry, representing the interconnectedness of all life. The pattern is believed to contain the blueprint of creation and has been found in numerous ancient cultures around the world.

- Fibonacci Sequence: The Fibonacci sequence is a mathematical series where each number is the sum of the two preceding numbers (0, 1, 1, 2, 3, 5, 8, 13, etc.). When represented geometrically, this sequence forms a spiral that mirrors the patterns found in nature, such as the arrangement of leaves on a stem or the shape of a seashell. This sequence is closely related to the golden ratio, which has been revered as the ideal proportion in both art and nature.

- Platonic Solids: The five Platonic solids—tetrahedron, hexahedron (cube), octahedron, dodecahedron, and icosahedron—are considered sacred because of their symmetrical properties. In sacred geometry, these shapes are thought to represent the building blocks of the physical universe, with each solid corresponding to one of the classical elements (earth, air, fire, water, and ether).

- Metatron's Cube: Metatron's Cube is a complex geometric figure that contains all five Platonic solids. In mystical traditions, it is believed to represent the structure of the universe, embodying the divine blueprint from which all things arise. Metatron's Cube is associated with the Archangel Metatron, who is said to oversee the flow of energy in creation.

Sacred Geometry and Christ Consciousness

Sacred geometry is not merely a set of abstract shapes and numbers; it is seen as the blueprint of divine consciousness itself. Christ Consciousness, which represents the divine awareness of unity, love, and interconnectedness, is expressed through these sacred geometric patterns. In this context, sacred geometry reveals the divine order and unity that underlie all of creation.
In many spiritual traditions, the act of creation is seen as a divine process governed by sacred geometric principles. John 1:1-3 tells us that "In the beginning was the Word, and the Word was with God, and the Word was God. Through him all things were made; without him, nothing was made that has been made." This "Word" can be understood as the primal vibration that brought the universe into existence, manifesting through sacred geometric patterns.

Christ Consciousness is the divine force that animates these patterns, infusing them with life and intelligence. Just as the sacred geometry of the Flower of Life represents the interconnectedness of all things, Christ Consciousness reflects the divine unity that connects every being in the cosmos. Sacred geometry provides a visual representation of the underlying order of the universe, which reflects the harmonious nature of Christ Consciousness. The balance and symmetry found in sacred geometric shapes mirror the balance and harmony inherent in divine consciousness. Just as Christ taught the importance of love, unity, and balance, these principles are encoded in the geometric structures of the universe.By meditating on sacred geometric patterns, individuals can attune themselves to Christ Consciousness and experience the unity of all creation. The contemplation of these shapes can serve as a spiritual practice that deepens one's awareness of the divine order and helps to awaken the higher consciousness that Christ exemplified.

The Universal Neural Network and Geometry

The concept of the universe as a giant neural network finds a strong parallel in sacred geometry. Just as neural networks in the brain consist of interconnected neurons that communicate through electrical signals, the universe consists of interconnected energy points that communicate through the vibrational patterns of sacred geometry. Sacred geometry provides a blueprint for understanding the structure of the universe and its energetic connections. For instance, the Flower of Life can be compared to a neural network, with each circle representing a node of consciousness and the overlapping circles symbolizing the interconnectedness of these nodes. Just as neural networks transmit information across synapses, sacred geometric patterns transmit divine intelligence across the cosmic web. In this framework, Christ Consciousness is the divine energy that flows through the geometric patterns of the universe, connecting all things in a vast network of consciousness. The sacred geometric shapes serve as pathways for this energy, guiding its flow and ensuring that the universe remains in harmony. By aligning oneself with these geometric patterns, an individual can connect to the universal neural network and become attuned to the flow of Christ Consciousness.

Practices for Awakening Christ Consciousness Through Sacred Geometry

There are several spiritual practices that can help individuals connect with the sacred geometry of the universe and awaken Christ Consciousness within themselves.

1. Meditation on Geometric Symbols: Meditating on sacred geometric shapes, such as the Flower of Life, the Sri Yantra, or Metatron's Cube, can help to align one's consciousness with the divine patterns of the universe. Visualizing these shapes during meditation allows one to attune to the vibrations of Christ Consciousness and experience a deeper sense of unity with all creation.

2. Sacred Art and Architecture: Many sacred sites around the world, such as cathedrals, temples, and pyramids, are built according to sacred geometric principles. Visiting these sites or creating art based on these patterns can help to activate Christ Consciousness. The contemplation of sacred art and architecture allows one to experience the divine harmony encoded in these structures.

3. Breathwork and Sound Healing: As discussed in Chapter 8, breath and sound are powerful tools for connecting with the geometric patterns of the universe. Practices such as Pranayama (breath control) or chanting sacred mantras (such as OM, which is considered the sound of creation) can help individuals resonate with the vibrational frequencies of sacred geometry and awaken Christ Consciousness within themselves.

Biblical References

1. Proverbs 3:19 – "By wisdom the Lord laid the earth's foundations, by understanding He set the heavens in place."
2. Job 38:4-5 – "Where were you when I laid the earth's foundation? Tell me, if you understand. Who marked off its dimensions? Surely you know!"
3. Colossians 1:16-17 – "For by Him all things were created: things in heaven and on earth, visible and invisible... He is before all things, and in Him all things hold together."
4. Isaiah 40:22 – "He sits enthroned above the circle of the earth, and its people are like grasshoppers. He stretches out the heavens like a canopy, and spreads them out like a tent to live in."

Chapter 4:
The Chakra System- Energy Vortices of the Human Body

In spiritual traditions, particularly in Hinduism, Buddhism, and various yogic systems, chakras are understood as energy centers within the human body. These energy vortices regulate physical, emotional, and spiritual aspects of an individual's being. Each chakra corresponds to specific bodily functions, states of consciousness, and spiritual experiences. By understanding and balancing the chakras, one can access deeper spiritual awareness, including the activation of Christ Consciousness.

Overview of the Seven Main Chakras

The traditional chakra system consists of seven primary energy centers that align with the spine, beginning at the base and rising to the crown of the head. Each chakra is associated with a specific color, element, and spiritual significance:

1. Root Chakra (Muladhara)
   - Location: Base of the spine
   - Color: Red
   - Element: Earth
   - Significance: The root chakra governs our sense of safety, security, and grounding. It connects us to the physical world and ensures that our basic needs for survival are met. When balanced, the root chakra fosters feelings of stability and confidence.

2. Sacral Chakra (Svadhisthana)
   - Location: Lower abdomen, just below the navel
   - Color: Orange
   - Element: Water
   - Significance: This chakra is the center of creativity, pleasure, and emotional balance. It governs sexual energy, relationships, and our capacity to experience joy. A balanced sacral chakra allows for healthy emotional expression and creative flow.

3. Solar Plexus Chakra (Manipura)
   - Location: Upper abdomen, near the diaphragm
   - Color: Yellow
   - Element: Fire

- Significance: The solar plexus chakra is the seat of personal power, self-confidence, and willpower. It regulates our ability to take action, assert ourselves, and manifest our desires. When aligned, this chakra enhances our sense of purpose and determination.

4. Heart Chakra (Anahata)
   - Location: Center of the chest
   - Color: Green
   - Element: Air
   - Significance: The heart chakra is the bridge between the physical and spiritual realms. It governs love, compassion, empathy, and forgiveness. When open and balanced, it fosters unconditional love for oneself and others, reflecting the principles of Christ Consciousness.

5. Throat Chakra (Vishuddha)
   - Location: Throat
   - Color: Blue
   - Element: Ether
   - Significance: The throat chakra governs communication, self-expression, and truth. It is the center through which we express our authentic selves and align our words with our higher truth. A balanced throat chakra allows for honest communication and creative expression.

6. Third Eye Chakra (Ajna)
   - Location: Forehead, between the eyes
   - Color: Indigo
   - Element: Light
   - Significance: The third eye chakra is the center of intuition, insight, and spiritual vision. It governs perception, imagination, and the ability to see beyond the material world. An open third eye chakra enhances spiritual clarity and wisdom.

7. Crown Chakra (Sahasrara)
   - Location: Top of the head
   - Color: Violet (or white)
   - Element: Cosmic Energy
   - Significance: The crown chakra connects us to the divine, representing spiritual enlightenment and the realization of unity consciousness. It is the gateway to higher states of awareness, including Christ Consciousness. When fully activated, it allows for the experience of oneness with the universe.

Christ Consciousness and Chakras

Christ Consciousness is the divine awareness of unity, love, and interconnectedness, and it flows through the chakra system. Each chakra represents a different aspect of this divine consciousness, and by aligning and balancing the chakras, one can awaken to higher levels of spiritual awareness. The flow of energy through the chakra system mirrors the flow of Christ Consciousness through the individual. This divine energy, often referred to as kundalini in yogic traditions, rises from the root chakra and ascends to the crown chakra, awakening spiritual awareness and deepening the experience of unity with the divine. In the context of Christ Consciousness, the chakras can be seen as vortices of divine energy, each representing a different aspect of Christ's teachings and consciousness.

For example:
- The heart chakra resonates deeply with Christ Consciousness because it is the center of love, compassion, and empathy. Christ's teachings on unconditional love are reflected in the energy of the heart chakra, which, when fully open, allows for the experience of divine love.
- The crown chakra represents the ultimate realization of Christ Consciousness—unity with the divine. As the chakra system reaches its culmination at the crown, the individual merges with universal consciousness, experiencing the oneness that Christ embodied.

Aligning the Chakras to Awaken Christ Consciousness

Balancing and aligning the chakras is essential for awakening Christ Consciousness. Various practices, including meditation, yoga, pranayama (breathwork), and energy healing techniques like Reiki (discussed in Chapter 5), can help to clear blockages and restore the free flow of divine energy through the chakra system. As energy flows freely from the root to the crown, the individual becomes more attuned to the divine principles of Christ Consciousness. This process is often referred to as a spiritual awakening or kundalini awakening in yogic traditions. During this awakening, the individual transcends the limitations of the ego and experiences a deep connection to the divine source of all life.

Chakras and the Neural Network of the Universe

The chakra system is not only a map of individual consciousness but also a reflection of the cosmic energy grid that permeates the universe. Just as neurons in the brain are interconnected through a vast network, the chakras are interconnected with the universal neural network of energy and consciousness. This network extends beyond the individual and connects all beings in a web of divine energy. Each chakra corresponds to a different aspect of the universal energy grid. For example, the root chakra aligns with the Earth's magnetic field, grounding the individual in the physical world, while the crown chakra aligns with the cosmic energy that connects all beings to the divine source. In this way, the chakras serve as points of connection between the individual and the universe, allowing for the flow of Christ Consciousness through both the microcosm (the individual) and the macrocosm (the cosmos). The energy that flows through the chakra system is the same energy that flows through the universe. Christ Consciousness represents the divine intelligence that permeates this energy grid, ensuring that all beings are interconnected in a web of love and unity. By aligning the chakras, individuals can tap into this universal flow of energy and experience the oneness that Christ embodied.

Practices for Balancing the Chakras and Awakening Christ Consciousness

Balancing and aligning the chakras is essential for awakening Christ Consciousness. Various practices, including meditation, yoga, pranayama (breathwork), and energy healing techniques like Reiki (discussed in Chapter 5), can help to clear blockages and restore the free flow of divine energy through the chakra system. As energy flows freely from the root to the crown, the individual becomes more attuned to the divine principles of Christ Consciousness. This process is often referred to as a spiritual awakening or kundalini awakening in yogic traditions. During this awakening, the individual transcends the limitations of the ego and experiences a deep connection to the divine source of all life.

There are several practices that can help individuals balance their chakras and awaken Christ Consciousness within themselves. These practices include meditation, energy healing, and sound therapy.

1. Chakra Meditation: Meditation is one of the most powerful tools for balancing the chakras and awakening Christ Consciousness. By focusing on each chakra and visualizing its corresponding color and energy, individuals can clear blockages and restore the free flow of divine energy. For example, visualizing a vibrant green light at the heart chakra can help to open the center of love and compassion, allowing for the experience of Christ's unconditional love.

2. Yoga and Breathwork (Pranayama): Yoga postures and breath control exercises are designed to stimulate the flow of energy through the chakra system. Specific postures, such as tree pose (for grounding the root chakra) or camel pose (for opening the heart chakra), can help to balance the chakras and awaken spiritual awareness. Pranayama, or breath control, is another powerful tool for clearing blockages in the chakra system and aligning oneself with Christ Consciousness.

3. Reiki and Energy Healing: As discussed in Chapter 5, Reiki is an energy healing practice that can be used to balance the chakras and attune oneself to the flow of Christ Consciousness. By channeling universal life force energy through the chakras, Reiki practitioners can help individuals to restore harmony and balance to their energy systems, facilitating spiritual awakening.

4. Sound Therapy: Each chakra resonates with a specific sound frequency, and using sound to balance the chakras can be a powerful way to awaken Christ Consciousness. For example, chanting the mantra OM can help to balance the crown chakra, while the sound LAM resonates with the root chakra. Tibetan singing bowls and tuning forks are also commonly used in sound therapy to align the chakras and promote spiritual healing.

Biblical References

1. Matthew 6:22 – "The eye is the lamp of the body. If your eyes are healthy, your whole body will be full of light." (Third Eye Chakra)
2. Proverbs 4:23 – "Above all else, guard your heart, for everything you do flows from it." (Heart Chakra)

3. Isaiah 40:31 – "But those who hope in the Lord will renew their strength. They will soar on wings like eagles; they will run and not grow weary, they will walk and not be faint." (Solar Plexus Chakra)

4. John 7:38 – "Whoever believes in me, as Scripture has said, rivers of living water will flow from within them." (Sacral Chakra)

Chapter 5: Reiki- Healing Through Universal Life Force

Reiki is a Japanese energy healing system that promotes the flow of universal life force energy through the body to heal physical, emotional, and spiritual ailments. The word Reiki is derived from two Japanese words: Rei meaning "universal" or "divine," and Ki meaning "life force energy." Reiki's foundational belief is that all living beings are imbued with this life force energy, which flows through and around them. When this energy is strong and balanced, individuals experience physical health, emotional well-being, and spiritual clarity. However, when the energy becomes blocked or stagnant, it can lead to illness and distress.

Reiki was developed by Mikao Usui in the early 20th century. Usui discovered this healing technique after a deep spiritual experience during a meditation retreat on Mount Kurama in Japan. The practice of Reiki involves the use of gentle touch or hand movements placed over the body to channel universal life force energy into the recipient. Reiki practitioners serve as conduits for this energy, allowing it to flow through their hands into the person receiving treatment, facilitating healing and restoring balance. Reiki is rooted in the idea that the body has the innate ability to heal itself when given the proper energetic support. This concept aligns with ancient healing traditions such as Ayurveda and Traditional Chinese Medicine (TCM), which also emphasize the importance of life force energy (prana in Ayurveda, qi in TCM) for maintaining health and well-being. At the core of Reiki lies the recognition of a divine, intelligent force that permeates all things. This life force energy is synonymous with the concept of Christ Consciousness, the divine awareness of unity, love, and healing that exists within and around all of creation. In Reiki, Christ Consciousness represents the ultimate source of healing energy that practitioners channel during sessions.

Christ Consciousness and Universal Life Force Energy

Christ Consciousness, as discussed in previous chapters, is the awareness of divine unity and love that transcends individual identity. It is the spiritual essence that connects all beings and manifests as unconditional love, compassion, and healing. In Reiki, this divine consciousness is experienced as universal life force energy, which flows through the practitioner and into the recipient, bringing about healing on all levels—physical, emotional, mental, and spiritual.
Reiki practitioners often describe the sensation of working with this energy as feeling a warm, tingling, or flowing sensation in their hands. This is the direct experience of Christ Consciousness moving through them, acting as a bridge between the divine source of healing and the person receiving treatment. By opening themselves to this divine flow, Reiki practitioners can facilitate profound healing and transformation.
In this way, Reiki can be seen as a practice that attunes individuals to the healing power of Christ Consciousness. Through the practice of Reiki, one can cultivate a deeper connection to this divine awareness and become a vessel for the expression of divine love and healing.

Practical Methods of Using Reiki to Attune to Christ Consciousness

Reiki provides practical methods for aligning oneself with Christ Consciousness and channeling this energy for personal healing and spiritual growth. Some of these methods include:

1. Self-Reiki: Practitioners can use Reiki on themselves by placing their hands over different parts of their body, focusing on areas where they feel tension, discomfort, or emotional imbalance. As they channel universal life force energy into these areas, they can set the intention to align with Christ Consciousness, allowing divine love and healing to flow through them.

2. Reiki Meditation: Reiki meditation involves using the principles of Reiki to clear and balance the energy centers (chakras) and attune oneself to the flow of divine energy. During this meditation, practitioners visualize Christ Consciousness as a bright, golden light that flows through their body, healing and restoring balance to their entire being.

3. Distance Reiki: Reiki can be performed from a distance, allowing practitioners to send healing energy to others across time and space. This practice reinforces the idea that Christ Consciousness is not limited by physical boundaries and that divine love and healing can be transmitted regardless of distance.

4. Reiki Symbols: Reiki practitioners use sacred symbols during sessions to enhance the flow of energy and connect more deeply with the divine source. These symbols, which include the Cho Ku Rei (Power Symbol) and Hon Sha Ze Sho Nen (Distance Healing Symbol), act as conduits for Christ Consciousness, amplifying the practitioner's ability to channel healing energy.

Reiki and the Human Electromagnetic Field

The human body generates an electromagnetic field (EMF), which surrounds and penetrates the physical body. This energy field, often referred to as the aura in spiritual traditions, reflects the health and vitality of the individual. A balanced, harmonious EMF is indicative of physical health, emotional well-being, and spiritual alignment, while a disturbed or imbalanced EMF may manifest as physical illness, emotional distress, or spiritual disconnection. Reiki works by restoring balance to the human EMF, allowing the life force energy to flow freely through the body and aura. When Reiki energy is channeled into the body, it helps to clear energetic blockages, strengthen the EMF, and realign the individual with the natural flow of universal life force energy. This process not only promotes physical healing but also fosters emotional release, mental clarity, and spiritual awakening. Christ Consciousness is the divine intelligence that underlies all creation, including the human EMF. When Reiki energy is channeled into the EMF, it infuses this field with the healing power of Christ Consciousness, bringing about a state of harmony and balance. As the EMF becomes more aligned with this divine energy, individuals experience greater physical vitality, emotional resilience, and spiritual clarity. Reiki practitioners often observe that, after receiving a session, individuals report feeling lighter, more centered, and deeply connected to their inner selves. This is because Reiki not only heals the physical body but also aligns the individual's energy field with the flow of Christ Consciousness, allowing them to experience a deeper sense of unity and peace.

Case Studies on Energy Healing and Spiritual Awakening Through Reiki

Numerous case studies have shown the profound impact of Reiki on both physical healing and spiritual awakening. Here are a few examples of how Reiki has facilitated transformation through the flow of Christ Consciousness:

1. Chronic Pain Relief: In one study, a woman suffering from chronic back pain received Reiki treatments over the course of several weeks. After each session, she reported a significant reduction in pain, improved mobility, and a sense of inner peace. By the end of the treatment period, her pain had diminished to the point where she no longer needed medication. Her practitioner attributed this healing to the flow of divine energy through her body, realigning her EMF and facilitating the release of physical tension.

2. Emotional Healing: A man experiencing severe depression and anxiety sought Reiki treatments as a complementary therapy to his traditional counseling. During his sessions, he felt a deep sense of warmth and comfort, which he described as "being held in the arms of God." Over time, he noticed a significant improvement in his emotional well-being, including reduced anxiety, greater self-compassion, and an increased ability to cope with stress. His Reiki practitioner explained that the energy of Christ Consciousness had helped to clear the emotional blockages in his EMF, allowing for emotional healing and spiritual awakening.

3. Spiritual Awakening: In another case, a woman who had been on a spiritual path for several years received Reiki treatments to deepen her connection to the divine. During one session, she experienced a powerful vision of light pouring through her crown chakra, filling her body with a sense of oneness with the universe. This experience marked a turning point in her spiritual journey, as she felt a deep connection to Christ Consciousness and a renewed sense of purpose in her life.

These examples illustrate the transformative power of Reiki as a tool for aligning the body, mind, and spirit with the flow of Christ Consciousness. Whether used for physical healing, emotional release, or spiritual awakening, Reiki provides a direct path to experiencing the divine love and unity that Christ Consciousness represents.

Biblical References

1. Matthew 10:8 – "Heal the sick, raise the dead, cleanse those who have leprosy, drive out demons. Freely you have received; freely give." (Christ Consciousness as a source of healing energy)
2. Mark 5:34 – "He said to her, 'Daughter, your faith has healed you. Go in peace and be freed from your suffering.'" (The power of faith in divine healing)

3. John 14:12 – "Very truly I tell you, whoever believes in me will do the works I have been doing, and they will do even greater things than these, because I am going to the Father." (Empowerment through Christ Consciousness)

4. James 5:15 – "And the prayer offered in faith will make the sick person well; the Lord will raise them up. If they have sinned, they will be forgiven." (Faith and divine healing)

Chapter 6: The Human Electromagnetic Field (EMF) as a Manifestation of Christ Consciousness

The human electromagnetic field (EMF) is a vital aspect of our existence that has gained attention from both scientific and spiritual communities. This field is composed of the electrical and magnetic energies generated by our bodies, particularly through the heart and brain. The EMF is often referred to as the aura, and it is believed to reflect the physical, emotional, and spiritual health of an individual. From a scientific standpoint, the human body generates electrical activity due to the functioning of the nervous system, muscle contractions, and cellular processes. This electrical activity creates a magnetic field that extends beyond the physical body, forming a complex, dynamic electromagnetic field. Research has shown that the heart produces the most powerful electromagnetic field in the body, which can be detected several feet away from the person. Studies utilizing electrocardiography (ECG) and electroencephalography (EEG) have demonstrated that the heart and brain communicate through electromagnetic signals. These signals are believed to influence our emotions, thoughts, and overall well-being. The EMF is also thought to play a role in our interactions with others, as it can be influenced by the energies of those around us.

Ancient Views on the Aura and the Energy Body

Ancient cultures have long recognized the existence of the human EMF. In traditional Chinese medicine, the aura is associated with the flow of qi, or life force energy, through the meridians of the body. Similarly, in Ayurveda, the energy body consists of various layers known as koshas, which include the physical, energetic, mental, emotional, and spiritual aspects of an individual. Spiritual traditions such as Hinduism and Buddhism also describe the aura in terms of chakras—energy centers that correspond to various aspects of our being. Each chakra is believed to be associated with specific qualities and functions, influencing both physical health and spiritual development.

The EMF and Christ Consciousness

Christ Consciousness, as previously discussed, is the awareness of divine unity and love that exists within all beings. This consciousness represents the ultimate state of being in alignment with the universal flow of energy and information. The human EMF serves as a manifestation of Christ Consciousness, reflecting our spiritual state and connection to the divine. The relationship between Christ Consciousness and the EMF can be understood as one of mutual influence. When individuals embody Christ Consciousness, their EMF becomes more vibrant, coherent, and expansive. Conversely, a balanced and harmonious EMF allows for the clearer expression of Christ Consciousness in one's life.

1. Vibrational Alignment: The EMF resonates with the frequencies of Christ Consciousness. When individuals engage in practices such as meditation, prayer, and energy healing, they raise their vibrational frequency, allowing their EMF to align with the divine energy of Christ Consciousness. This alignment enhances their ability to connect with the universe and receive divine guidance.

2. Healing and Restoration: The flow of Christ Consciousness through the EMF can facilitate profound healing and restoration. When the EMF is balanced, it allows for the free flow of energy, promoting physical health and emotional well-being. In contrast, a disrupted EMF may lead to illness, emotional disturbances, and spiritual disconnection.

3. Awakening Higher States of Awareness: As individuals cultivate a deeper connection with Christ Consciousness, their EMF expands, enabling them to experience higher states of awareness. This expansion can lead to increased intuition, creativity, and a greater sense of interconnectedness with others and the universe.

Practices for Cultivating a Balanced and Strong EMF through Spiritual Practices

Cultivating a balanced and strong EMF is essential for maintaining overall health and enhancing one's connection to Christ Consciousness. Here are several practices that can help individuals strengthen their EMF:

1. Meditation: Regular meditation helps to calm the mind, reduce stress, and enhance the flow of energy through the body. It allows individuals to connect with their inner selves and align with Christ Consciousness.

2. Breathwork: Conscious breathing techniques can help to increase energy flow and harmonize the EMF. Practices such as pranayama in yoga or Holotropic Breathwork can facilitate a deeper connection to the life force energy.

3. Grounding: Grounding, or earthing, involves connecting with the Earth's energy by walking barefoot on natural surfaces, such as grass, soil, or sand. This practice helps to stabilize the EMF and promotes feelings of peace and balance.

4. Energy Healing: Engaging in energy healing modalities such as Reiki, Qi Gong, or Tai Chi can help to clear energetic blockages, restore balance, and enhance the flow of Christ Consciousness within the EMF.

5. Nature Connection: Spending time in nature allows individuals to attune their EMF to the natural rhythms of the Earth. Activities such as hiking, gardening, or simply sitting in a park can help to recharge and revitalize the energy field.

The EMF as a Neural Network

The human EMF can be viewed as a microcosm of the larger universal neural network. Just as the brain is composed of interconnected neurons that transmit information, the human EMF serves as a network of energy that communicates with the larger cosmic field. This perspective highlights the interconnectedness of all beings and the flow of information and energy through the universe.

1. Information Exchange: The EMF acts as a conduit for information exchange between individuals and the universe. When individuals are in tune with their EMF, they can access higher states of consciousness and receive insights and guidance from the divine.

2. Collective Consciousness: The human EMF contributes to the collective consciousness of humanity. As individuals align their EMF with Christ Consciousness, they strengthen the collective energy field, promoting healing and unity on a global scale.

3. Interconnectedness: The EMF serves as a reminder of the interconnectedness of all beings. Each individual's energy field contributes to the larger cosmic web, emphasizing the importance of maintaining a balanced and harmonious EMF for the benefit of the entire universe.

Biblical References

1. 1 Corinthians 12:12 – "For just as the body is one and has many members, and all the members of the body, though many, are one body, so it is with Christ." (Emphasizing interconnectedness and unity)
2. Romans 12:5 – "So we, though many, are one body in Christ, and individually members one of another." (Highlighting the collective energy field of humanity)
3. Ephesians 4:16 – "From whom the whole body, joined and knit together by what every joint supplies, according to the effective working by which every part does its share, causes growth of the body for the edifying of itself in love." (The importance of each individual's energy contribution)
4. Colossians 1:17 – "He is before all things, and in Him all things hold together." (Christ Consciousness as the glue that holds the universe together)

Chapter 7: The Electromagnetic Field of the Natural Environment

The Earth's electromagnetic field (EMF) is an intricate system that plays a critical role in maintaining the balance of life on our planet. This field is not merely a scientific phenomenon; it resonates deeply with spiritual principles and offers insights into the interconnectedness of all beings. The Earth's magnetic field is generated by the movement of molten iron in its outer core. This magnetic field extends into space and protects the planet from harmful solar radiation and cosmic rays. It is a vital aspect of the Earth's environment, influencing everything from navigation to weather patterns. Understanding the relationship between the human EMF and the Earth's electromagnetic field is crucial for promoting harmony and balance in both individual and collective experiences. This connection emphasizes the importance of being attuned to the natural world and recognizing our role within the greater cosmic system. Spiritually, many cultures have recognized the significance of the Earth's magnetic field as a manifestation of divine energy. Indigenous traditions often speak of the Earth as a living entity, embodying the wisdom and spirit of nature. This concept is closely aligned with the idea of Gaia theory, which posits that the Earth functions as a single, self-regulating organism.

1. Spiritual Traditions: Various spiritual traditions emphasize the importance of connecting with the Earth's energy. For instance, in Native American culture, the Earth is viewed as sacred, and rituals are performed to honor its spirit. Similarly, in Hinduism, the Earth is considered Bhumi Devi, the goddess who sustains life. Such traditions recognize the interdependence of all beings and the necessity of respecting and nurturing the Earth's energy.

2. The Role of Ley Lines: Ley lines, often described as invisible pathways of energy that traverse the Earth, are thought to connect sacred sites and natural features. These lines represent the flow of electromagnetic energy and have been revered in various cultures as sources of spiritual power. Aligning with these energy lines can enhance one's connection to Christ Consciousness and foster healing.

The presence of Christ Consciousness is not confined to human experience; it permeates the natural world, reflecting the divine order and interconnectedness of all life. This chapter explores how nature serves as a mirror of divine consciousness and a source of inspiration for spiritual growth.

How Natural Environments Serve as Mirrors of Divine Consciousness

1. Nature as a Teacher: Nature has long been regarded as a profound teacher of spiritual wisdom. Observing the cycles of life, the interdependence of ecosystems, and the beauty of

natural phenomena can inspire individuals to reflect on their place in the universe and the importance of harmony. The rhythms of nature mirror the divine flow of Christ Consciousness, encouraging us to align with these natural patterns.

2. Healing Properties of Nature: Numerous studies highlight the healing benefits of spending time in natural environments. Research has shown that exposure to nature can reduce stress, improve mood, and enhance overall well-being. This healing aspect of nature is often attributed to its grounding effect, connecting individuals to the Earth's energy and facilitating the flow of Christ Consciousness.

3. Sacred Sites and Energy Centers: Certain locations on Earth are considered sacred, often linked to powerful energies that resonate with Christ Consciousness. Sites such as the pyramids of Egypt, Stonehenge, and the sacred mountains of Tibet are thought to serve as energy centers, amplifying the flow of divine consciousness. Pilgrimages to these sites can deepen one's spiritual connection and facilitate healing.

The Human-Environment Connection

1. Resonance and Synchronization: Human EMFs are influenced by the Earth's magnetic field, creating a symbiotic relationship between individuals and their environment. When individuals are in harmony with the Earth's energy, their EMFs resonate with the planetary field, enhancing their spiritual experience and fostering a sense of unity.

2. Electromagnetic Pollution: In today's modern world, exposure to artificial electromagnetic fields from technology can disrupt the natural balance of the human EMF. This disruption can lead to stress, anxiety, and various health issues. Cultivating awareness of this interaction and making conscious choices to reduce exposure to harmful EMFs can promote overall well-being and strengthen the connection with Christ Consciousness.

3. Sustainable Practices: Living in harmony with the Earth's EMF involves adopting sustainable practices that honor and protect the environment. By engaging in eco-friendly practices such as reducing waste, conserving energy, and supporting sustainable agriculture, individuals can foster a positive relationship with the Earth and align with the principles of Christ Consciousness.

Biblical References

1. Psalm 19:1 – "The heavens declare the glory of God; the skies proclaim the work of his hands." (Nature as a reflection of divine glory)
2. Genesis 1:31 – "God saw all that he had made, and it was very good." (Recognition of the goodness of creation).

3. Romans 1:20 – "For since the creation of the world God's invisible qualities—his eternal power and divine nature—have been clearly seen, being understood from what has been made." (The divine nature reflected in creation).

4. Isaiah 55:12 – "You will go out in joy and be led forth in peace; the mountains and hills will burst into song before you, and all the trees of the field will clap their hands." (Nature rejoicing in divine presence).

Chapter 8: The Power of Sound: Vibrations, Healing, and Christ Consciousness

Sound has long been recognized as a fundamental aspect of both the physical and spiritual worlds. This chapter explores the powerful role that sound plays in connecting individuals to Christ Consciousness, the principles of vibrational healing, and the ways in which sound can facilitate spiritual awakening and well-being. In many spiritual traditions, the concept of the "Word" symbolizes the divine creative force that brings forth creation. In Christianity, the opening verse of the Gospel of John states: John 1:1-3 "In the beginning was the Word, and the Word was with God, and the Word was God. He was with God in the beginning. Through him all things were made; without him nothing was made that has been made." This passage illustrates the fundamental role of sound as the primal vibration that underlies all existence. Christ Consciousness, as the embodiment of divine love and unity, resonates with this creative force, inviting individuals to align with it through their own vocalizations and intentions.

Sound Healing in Ancient and Modern Traditions

Sound healing is an ancient practice found in many cultures around the world, each with its unique approaches and techniques. This section will explore a few of the most notable sound healing traditions, highlighting their principles and practices.

1. Tibetan Singing Bowls: These bowls, made from a blend of metals, produce rich harmonic tones when struck or circled with a mallet. Tibetan singing bowls have been used in Buddhist traditions for centuries to promote meditation, relaxation, and healing. The vibrations produced by the bowls resonate with the body's energy centers, facilitating the release of blockages and the flow of healing energy.

2. Mantras and Chanting: In many spiritual traditions, vocalization of sacred sounds, or mantras, is considered a powerful tool for aligning with higher consciousness. In Hinduism and Buddhism, chanting mantras such as "Om" creates vibrations that resonate with the divine and elevate the practitioner's consciousness. This practice can open pathways to experiencing Christ Consciousness through the resonance of sound.

3. Shamanic Drumming: Shamanic traditions utilize rhythmic drumming to induce altered states of consciousness and connect with the spirit world. The repetitive beats of the drum serve as a bridge between the physical and spiritual realms, facilitating healing journeys and enhancing spiritual insight. This practice exemplifies how sound can act as a conduit for divine energies.

4. Native American Flute Music: The Native American flute is often used in spiritual ceremonies and healing practices. The gentle melodies produced by the flute evoke a sense of peace and connection to nature, allowing listeners to tap into the vibrational energy of the Earth and the

universe. This music reflects the understanding that sound can carry the essence of Christ Consciousness.

The Science of Sound Frequencies and Their Effects on the Body and Mind

Recent advancements in scientific research have shed light on the profound effects of sound frequencies on human health and consciousness.

1. Frequency and Healing: Every object, including the human body, has its own vibrational frequency. When the body is exposed to sound frequencies that align with its natural rhythm, it can promote healing and balance. Research has shown that certain frequencies can reduce stress, alleviate pain, and enhance overall well-being. For instance, the Solfeggio frequencies, which range from 396 Hz to 963 Hz, are believed to have specific healing properties that resonate with different aspects of the human experience.

2. Brainwave Entrainment: Sound has the power to influence brainwave patterns, a phenomenon known as brainwave entrainment. By exposing the brain to specific frequencies, individuals can achieve states of relaxation, meditation, or heightened awareness. This process facilitates access to deeper levels of consciousness, allowing for the experience of Christ Consciousness.

3. Vibrational Medicine: Sound therapy is increasingly recognized as a form of vibrational medicine, utilizing specific sound frequencies to promote healing. Techniques such as binaural beats and sound baths are employed to create immersive sound experiences that align the body and mind with healing vibrations. These practices facilitate the release of emotional blockages, allowing individuals to connect with their higher selves and divine consciousness.

Sound as a Carrier of Divine Energy and Christ Consciousness

1. Prayer and Intention: The act of prayer, often spoken aloud, can serve as a powerful means of connecting with Christ Consciousness. When individuals vocalize their intentions, they create sound vibrations that carry their desires into the universe. This practice amplifies their alignment with divine energy and fosters a deeper connection to the Christ Consciousness within.

2. Musical Expression: Music has the ability to transcend language and evoke deep emotional responses. Many musicians and composers draw inspiration from spiritual themes, channeling divine energies through their art. By listening to or participating in musical expressions that resonate with Christ Consciousness, individuals can experience healing and upliftment.

3. Sonic Meditations: Guided meditations that incorporate sound, such as crystal bowls or chimes, can facilitate deep states of relaxation and awareness. These sonic meditations allow individuals to tap into the vibrations of Christ Consciousness, promoting healing, insight, and spiritual growth.

Sound and the Universal Neural Network- How Sound Waves Resonate with the Universal Network of Energy and Consciousness

1. Interconnected Vibrations: Sound waves travel through the universe, creating ripples of energy that connect all beings. This interconnectedness reflects the neural network model, where each sound vibration acts as a synapse in the larger system of consciousness. By recognizing the significance of sound in the universal energy grid, individuals can become more attuned to the flow of Christ Consciousness.

2. Healing Through Resonance: As sound waves resonate with the body's energy centers, they can facilitate healing and transformation. The practice of using sound healing in conjunction with spiritual principles enhances the ability to connect with divine energies. This connection fosters a deeper understanding of the interdependence of all beings and the importance of living in harmony with the universe.

3. Enhancing Spiritual Practices: Incorporating sound into spiritual practices, such as meditation or yoga, can deepen the experience of Christ Consciousness. The vibrations of sound help create a conducive environment for inner exploration and connection with the divine. This alignment allows individuals to access higher states of awareness and fosters personal and collective healing.

Biblical References

1. Psalm 150:6 – "Let everything that has breath praise the Lord." (The call to use sound in worship and connection to the divine)
2. 1 Samuel 16:23 – "Whenever the spirit from God came on Saul, David would take up his lyre and play. Then relief would come to Saul; he would feel better, and the evil spirit would leave him." (The healing power of music)
3. Revelation 14:2 – "And I heard a sound from heaven like the roar of rushing waters and like a loud peal of thunder." (The divine sound resonating with power)
4. Matthew 18:20 – "For where two or three gather in my name, there am I with them." (The collective sound of voices uniting in Christ)

Chapter 9: Alternative Medicines from Around the World: Christ Consciousness and Healing

In this chapter, we will explore the diverse range of alternative healing practices that exist globally, each of which embodies principles of Christ Consciousness. We will investigate how these traditions promote healing through a deep connection to divine energy and the innate healing capacities of individuals. By integrating insights from these alternative medicine systems with modern scientific understanding, we can appreciate the holistic approach to health and well-being.

Global Healing Traditions

1. Ayurveda: This ancient system of medicine from India emphasizes balance among the body's three doshas—Vata, Pitta, and Kapha. Ayurveda views health as a state of equilibrium between the body, mind, and spirit. It employs natural remedies, dietary practices, and holistic lifestyle choices to restore harmony. The understanding of Christ Consciousness is intrinsic to Ayurveda, as practitioners recognize the divine presence in all aspects of life and the importance of connecting with this energy for optimal health.

2. Traditional Chinese Medicine (TCM): TCM is based on the principles of balance and harmony within the body and with nature. It employs modalities such as acupuncture, herbal medicine, and qi gong to restore the flow of chi (life force energy). TCM acknowledges the interconnectedness of all beings and the significance of aligning with universal energies, echoing the teachings of Christ Consciousness.

3. Indigenous Healing Systems: Indigenous cultures worldwide have their unique healing practices that reflect their spiritual beliefs and connection to the Earth. These systems often involve herbal medicine, shamanic rituals, and ceremonies that honor the spirit of nature. The emphasis on community and the collective well-being of all beings resonates with the principles of Christ Consciousness, promoting healing through shared love and connection.

4. Homeopathy: Developed in the 18th century by Samuel Hahnemann, homeopathy is based on the principle of "like cures like." It utilizes highly diluted substances to stimulate the body's innate healing responses. Homeopathy recognizes the importance of treating the individual as a whole, considering physical, emotional, and spiritual aspects. The practice reflects the Christ Consciousness ethos of addressing the root cause of illness rather than just symptoms.

5. Naturopathy: Naturopathy focuses on supporting the body's natural healing abilities through holistic practices, including nutrition, herbal medicine, and lifestyle changes. Naturopathic practitioners view health as a dynamic balance of physical, emotional, and spiritual well-being,

embodying the principles of Christ Consciousness by encouraging individuals to connect with their inner healing potential.

Christ Consciousness as a Healing Force- Manifestations of Divine Energy in Healing Practices

1. Healing Intentions: In many alternative healing modalities, practitioners focus on setting positive intentions for their clients. This practice aligns with the energy of Christ Consciousness, emphasizing love, compassion, and the desire for holistic healing. By channeling this divine energy, practitioners can create a conducive environment for healing and transformation.

2. Energy Work: Various alternative healing practices, such as Reiki and healing touch, involve the manipulation of energy to promote healing. These modalities recognize the existence of a universal life force that flows through all living beings, echoing the teachings of Christ Consciousness. Practitioners learn to attune themselves to this energy, facilitating the healing process for themselves and others.

3. Meditation and Mindfulness: Many alternative healing systems incorporate meditation and mindfulness as integral components. These practices help individuals cultivate inner peace, connect with their higher selves, and align with Christ Consciousness. By quieting the mind and focusing on the present moment, individuals can access deeper states of awareness and healing.

4. Spiritual Guidance: Alternative medicine practitioners often integrate spiritual guidance into their work, drawing upon divine wisdom to inform their healing practices. This guidance can take various forms, such as intuitive insights, prayers, or channeling. The presence of Christ Consciousness in these interactions fosters an environment of trust and love, promoting healing on all levels.

Integration of Alternative Medicine with Modern Science- The Holistic View of Health

1. Mind-Body Connection: Modern scientific research has increasingly acknowledged the mind-body connection in health and healing. Studies have demonstrated that emotional and psychological states can significantly impact physical health. Alternative medicine approaches, with their holistic perspectives, align with this understanding by addressing the interplay between mental, emotional, and physical well-being.

2. Research on Energy Healing: Several studies have explored the efficacy of energy healing modalities, such as Reiki and therapeutic touch. Research findings indicate that these practices can reduce stress, alleviate pain, and promote relaxation. These results support the notion that Christ Consciousness, as a universal healing force, can facilitate physical and emotional healing.

3. Integrative Health Approaches: The integration of alternative medicine with conventional healthcare is gaining recognition. Many healthcare practitioners are now embracing holistic approaches, incorporating alternative therapies alongside traditional treatments. This integration reflects an understanding of the multifaceted nature of health and well-being, resonating with the principles of Christ Consciousness by promoting comprehensive healing.

Biblical References

1. Jeremiah 30:17 – "But I will restore you to health and heal your wounds,' declares the Lord." (God's promise of healing)
2. James 5:14-15 – "Is anyone among you sick? Let them call the elders of the church to pray over them and anoint them with oil in the name of the Lord. And the prayer offered in faith will make the sick person well; the Lord will raise them up." (The power of prayer and community in healing).
3. Exodus 15:26 – "I am the Lord who heals you." (Affirmation of divine healing).
4. Matthew 9:35 – "Jesus went through all the towns and villages, teaching in their synagogues, proclaiming the good news of the kingdom and healing every disease and sickness." (Christ as a healer).

Chapter 10: The Journey to Awakening: Embodying Christ Consciousness in Everyday Life

In this final chapter, we will delve into practical strategies for awakening and embodying Christ Consciousness in our daily lives. This journey is not merely about intellectual understanding; it requires active engagement and conscious effort to align ourselves with the divine principles of love, unity, and compassion. By adopting daily practices that resonate with Christ Consciousness, we can foster a deeper connection with ourselves, others, and the universe. The journey to awakening Christ Consciousness is a lifelong endeavor that requires commitment, intention, and love.

Daily Practices for Awakening Christ Consciousness

1. Meditation and Contemplation

Meditation is a powerful tool for connecting with Christ Consciousness. By creating a quiet space within, we can open ourselves to divine wisdom and insight. Here are some meditation practices to consider:

- Mindfulness Meditation: Focus on your breath and bring your attention to the present moment. Allow thoughts to arise without judgment, acknowledging them before returning to your breath. This practice cultivates awareness and helps you connect with the present moment, the essence of Christ Consciousness.

- Loving-Kindness Meditation (Metta): Begin by cultivating love and compassion for yourself, then gradually extend this feeling to others, including friends, family, and even those with whom you may have conflicts. This practice aligns with Christ's teachings of love and compassion.

- Guided Visualization: Visualize yourself surrounded by divine light, experiencing the warmth and love of Christ Consciousness. Allow this energy to fill you and radiate outward to all beings.

2. Breathwork and Energy Practices

Breath is a vital link between the physical and spiritual realms. Engaging in breathwork can help activate Christ Consciousness within us.

- Deep Breathing Exercises: Practice deep, diaphragmatic breathing to calm the mind and energize the body. Inhale deeply through your nose, hold for a moment, and exhale slowly through your mouth. Visualize each breath drawing in divine energy and exhaling any negativity.

- Pranayama: This ancient yogic practice focuses on controlling breath to balance the body's energy. Techniques such as Nadi Shodhana (alternate nostril breathing) can help harmonize your energies and connect you with higher states of consciousness.

3. Energy Healing and Reiki

Practicing energy healing modalities like Reiki can facilitate your alignment with Christ Consciousness. Here are ways to incorporate energy work into your daily routine:

- Self-Reiki: Set aside time daily to practice self-Reiki. Place your hands on various energy centers in your body, focusing on sending love and healing energy to yourself. Visualize Christ Consciousness flowing through you, promoting healing and awakening.

- Healing Touch: If you have trained in any energy healing techniques, offer sessions to friends, family, or pets. As you channel healing energy, intend to connect with Christ Consciousness and invite this divine love into your work.

4. Journaling and Reflection

Writing can be a transformative practice to deepen your connection to Christ Consciousness.

- Daily Gratitude Journaling: Each day, write down three things for which you are grateful. This practice cultivates a mindset of abundance and appreciation, aligning you with the loving energy of Christ.

- Reflective Journaling: Take time to reflect on your thoughts, feelings, and experiences. Ask yourself how you can embody Christ Consciousness more fully in your life and what steps you can take toward alignment.

5. Acts of Service and Compassion

Embodying Christ Consciousness involves actively serving others and spreading love in the world.

- Volunteering: Engage in community service or volunteer at local organizations. Acts of kindness and service resonate deeply with Christ's teachings and help awaken the Christ Consciousness within you.

- Random Acts of Kindness: Incorporate small acts of kindness into your daily life, such as complimenting a stranger, helping someone in need, or simply offering a smile. These gestures create ripples of love and compassion in the world.

6. Living with Intention and Awareness

To embody Christ Consciousness, it's essential to live with intention and mindfulness.

- Set Daily Intentions: Each morning, set a clear intention for how you wish to embody Christ Consciousness that day. This intention can guide your actions and interactions, aligning you with divine purpose.

- Practice Mindful Presence: Throughout the day, practice being present in each moment. Whether you're eating, walking, or conversing, immerse yourself fully in the experience, allowing the presence of Christ Consciousness to guide you.

Living in Alignment with the Universal Neural Network

To live in harmony with the energetic patterns of the universe is to recognize that we are all interconnected and part of a larger divine tapestry. Here are some ways to deepen this connection:

- Nature Connection: Spend time in nature, observing and appreciating the beauty of creation. Recognize the divine energy present in all living things, fostering a sense of unity with the Earth and its inhabitants.

- Mindful Consumption: Be conscious of what you consume, both physically and energetically. Choose organic and ethically sourced foods, and limit exposure to negative media. Consider how your choices impact the world and strive to live in a way that reflects Christ Consciousness.

- Cultivating Community: Surround yourself with like-minded individuals who share your commitment to awakening Christ Consciousness. Engage in group meditations, discussions, or workshops that promote spiritual growth and collective healing.

Glossary

1. Alternative Medicine
A broad category of healing practices that fall outside conventional Western medicine. These include Ayurveda, Traditional Chinese Medicine, Indigenous healing traditions, and Reiki, often focusing on holistic health and energy balance.

2. Aura
The human energy field that surrounds the body, often associated with electromagnetic energy. The aura is believed to reflect a person's spiritual, emotional, and physical health and can be influenced by practices such as meditation and energy healing.

3. Biofield
Another term for the human electromagnetic field (EMF), which surrounds and interpenetrates the body. It reflects the state of one's physical, emotional, and spiritual health and can be influenced by energy healing practices.

4. Chakras
Energy centers within the human body according to Eastern spiritual traditions. The seven primary chakras run along the spine, from the root (base) to the crown (top of the head), each governing different aspects of physical, emotional, and spiritual well-being.

5. Christ Consciousness
A state of heightened spiritual awareness embodying divine love, unity, and compassion. It represents the realization of the interconnectedness of all beings, transcending religious boundaries, and reflecting the universal teachings of Christ about oneness with the divine.

6. Divine Intelligence
A concept that suggests a higher, conscious force governs the universe. In the context of Christ Consciousness, divine intelligence is the creative and sustaining force of love, unity, and order that flows through all life.

7. Electromagnetic Field (EMF)
A field produced by electrically charged objects, which exerts force on other charges in the vicinity. In the human body, the EMF is believed to be an energetic manifestation of health and consciousness, often referred to as the aura or biofield.

8. Electromagnetism
A fundamental force of nature associated with electrically charged particles. It governs phenomena such as light, radio waves, and magnetic fields, and plays a crucial role in understanding both the physical universe and metaphysical concepts like energy fields.

9. Energy Vibration

A fundamental concept in quantum physics and spiritual traditions, referring to the frequency at which energy moves. Higher vibrations are associated with spiritual growth, love, and positivity, while lower vibrations align with fear, negativity, and stagnation.

10. Ether
In ancient philosophy and metaphysical thought, ether (or aether) is described as the fifth element, a subtle, invisible medium that permeates the universe, facilitating the flow of energy and spiritual forces, and connecting the physical and spiritual realms.

11. Fibonacci Sequence
A mathematical sequence where each number is the sum of the two preceding ones. Found in nature (e.g., in the growth patterns of plants), the sequence is seen as a reflection of the divine order of the universe, often linked to sacred geometry.

12. Flower of Life
A geometric figure composed of overlapping circles arranged in a hexagonal pattern. It is regarded as a symbol of creation, representing the fundamental patterns of life and the interconnectedness of all beings.

13. Frequency
The rate at which a wave or vibration occurs, often linked to energy and consciousness. In spirituality, different frequencies are believed to resonate with different states of consciousness, healing, and spiritual insight.

14. Gaia Consciousness
The belief that the Earth (Gaia) is not only a physical entity but also possesses consciousness. This consciousness is interwoven with Christ Consciousness, reflecting the Earth's role in nurturing and sustaining all forms of life.

15. Gaia Theory
A scientific and spiritual concept that views the Earth as a living, self-regulating organism. This theory aligns with the belief that the Earth has its own consciousness and that its systems work together in a delicate balance, often seen as part of Christ Consciousness.

16. Mantra
A sacred sound, word, or phrase repeated in meditation to invoke spiritual energy or a state of higher consciousness. In many traditions, mantras are believed to connect the individual with divine consciousness or Christ Consciousness.

17. Nadis
In Eastern traditions, nadis are the channels through which prana flows within the body. They are akin to meridians in Traditional Chinese Medicine, and maintaining a clear flow of prana is essential for health and spiritual awakening.

18. Platonic Solids
Geometric shapes named after the philosopher Plato, considered the building blocks of the universe. These shapes, such as the cube and the dodecahedron, are associated with different elements and spiritual forces in sacred geometry.

19. Prana
A concept from Hindu philosophy referring to the vital life force that animates all living beings. Similar to the Chinese concept of "Qi," prana flows through the body's energy channels (nadis) and is integral to balancing physical and spiritual health.

20. Quantum Entanglement
A phenomenon in quantum physics where particles remain connected across vast distances, such that the state of one instantly influences the state of another. It serves as a scientific parallel to spiritual beliefs about the interconnectedness of all things in Christ Consciousness.

21. Quantum Field Theory
A branch of physics that describes the universe as a collection of quantum fields, with energy and particles constantly fluctuating within them. It forms a modern scientific parallel to ancient concepts of ether and interconnectedness.

22. Reiki
A Japanese energy healing technique that involves the transfer of universal life force energy through the practitioner's hands to the recipient. Reiki is based on the idea that this energy flows through all living things and can heal and balance the mind, body, and spirit.

23. Sacred Geometry
The study of geometric shapes and patterns that are believed to have inherent spiritual significance. These shapes, such as the Flower of Life and the Fibonacci sequence, are seen as blueprints of creation, representing the unity and harmony of the cosmos.

24. Shamanic Drumming
A traditional practice where rhythmic drumming is used to induce a meditative or trance-like state. In many cultures, it is believed that the drum's vibrations connect the individual with spiritual realms and universal consciousness.

25. Sound Frequencies
The measurable vibration of sound waves, typically described in Hertz (Hz). In spiritual practices, certain sound frequencies are believed to align with different chakras and states of consciousness, facilitating healing and spiritual connection.

26. Sound Healing
The use of sound frequencies to promote healing and balance in the body and mind. Practices include Tibetan singing bowls, mantras, and drumming, all of which are believed to resonate with the energy body and Christ Consciousness to facilitate spiritual awakening and healing.

27. Spiritual Awakening
The process of becoming more aware of the spiritual dimensions of life. It often involves experiencing Christ Consciousness, realizing the interconnectedness of all beings, and aligning with the higher purposes of love, compassion, and unity.

28. Unified Field Theory
A theoretical framework in physics that attempts to unite all fundamental forces of nature (gravity, electromagnetism, weak and strong nuclear forces) under one cohesive theory. Spiritually, it aligns with the idea that all forces are expressions of divine consciousness.

29. Universal Neural Network
A metaphorical concept that likens the universe to a giant neural network, where all matter and energy are interconnected, similar to the way neurons in the brain are connected. It reflects the interconnectedness of all beings, energy, and consciousness.

30. Vortex
A spiraling flow of energy that can occur in nature, the body, or even the cosmos. In the context of chakras, a vortex refers to the spiral motion of energy within each chakra, connecting the individual to the flow of Christ Consciousness.

References

1. Hahnemann, S. (1996). The Organon of Medicine. (Trans. from the 6th German edition). New Delhi: B. Jain Publishers.
2. Lu, H. (2018). The Role of Energy in Healing: Reiki, Therapeutic Touch, and their Influence on Health. Journal of Alternative and Complementary Medicine, 24(4), 317-322.
3. Young, D. (2015). Ayurveda: A Comprehensive Guide to Traditional Indian Medicine for the West. New York: Healing Arts Press.
4. Ma, Y., & Feng, X. (2019). The effect of Qi Gong on improving the quality of life and mental health of patients with chronic diseases: A systematic review. Journal of Psychosomatic Research, 125, 109786.
5. Van Oudenhove, L., & De Witte, K. (2018). Integrative medicine and its role in contemporary health care. Journal of Holistic Nursing, 36(1), 54-64.